iPhone 16 User

The Ultimate iPhone 16 Companion Glossary, Icon Index, iOS 18 Shortcuts & Support Guide for Beginners, Seniors, and Advanced Users—Everything You Need to Master Your iPhone with Confidence

Georgette Howard

Table of Contents

7

effort has been made to ensure accuracy, the author and publisher assume no responsibility for any errors, omissions, or outcomes resulting from the use of this guide. Always refer to Apple's official documentation or customer support for the most up-to-date and accurate information.

INTRODUCTION

Why This Book Is Different

Welcome to the *iPhone 16 Made Easy experience*, a book born from real questions, everyday frustrations, and the universal desire to use technology *without feeling overwhelmed.*

This isn't just another technical manual. It's a **step-by-step companion** built for the real world. Whether you're holding your very first iPhone, upgrading from an older model, or simply want to feel more confident navigating your device—this book was designed for **you**.

Inside, you'll find plain-English explanations, clear visuals, simple instructions, and pro tips that make even the most advanced iPhone 16 features easy to master. We've stripped away the jargon, skipped the fluff, and focused entirely on **what matters most to you**.

No confusing tech talk. No assuming you already know

how. Just pure, human-friendly help.

Who This Book Is For

This guide was carefully crafted for:

- ✓ **Seniors & Non-Tech Users** who want gentle, easy-to-follow support.
- ✓ **Everyday Beginners** who just want to call, text, and take great photos.
- ✓ **Dummies (Newbies)** who feel left behind by tech jargon.
- ✓ **Busy Professionals** looking to optimize their iPhone for work.
- ✓ **iPhone Creatives & Power Users** who want to unlock expert-level tricks.

Whether you're 18 or 88, if you want to **stop struggling and start enjoying** your iPhone 16—this book is for you.

What You'll Learn

By the end of this book, you'll be able to:

- ✓ Set up your iPhone 16 from scratch—*without needing help from anyone.*
- ✓ Make calls, send messages, and FaceTime with ease.
- ✓ Master the camera for pro-level photos and videos.
- ✓ Navigate iOS 18 like a pro, using gestures and hidden features.
- ✓ Stay connected with email, social apps, Safari, and iCloud.
- ✓ Use Siri and Shortcuts to automate your day.
- ✓ Customize your screen, widgets, Focus mode, and privacy settings.
- ✓ Fix common problems quickly without calling Apple support.
- ✓ Use secret iPhone tips, tricks, and expert-level hacks most users never discover.

And best of all, you'll finally **feel confident, in control,** and **empowered** with your iPhone—whether you're calling a friend, capturing a memory, or simply getting things done.

How to Use This Book

This isn't a book you need to read cover-to-cover.

Instead, treat it like a **friendly guidebook** you can jump into any time. Each chapter covers a specific part of your iPhone—like calling, camera, settings, or privacy and includes:

- **Screenshots** so you can see exactly what to tap or swipe.
- **Checklists** for quick wins and confidence.
- **Tips & Tricks Boxes** to save time or reveal cool features.
- **Troubleshooting Fixes** to solve common frustrations fast.
- **For Seniors / For Pros Callouts** to guide your experience based on your level.

Look for these helpful icons throughout the book:

- 👓 = Senior-friendly tip
- 💼 = Pro-level insight
- ⚠️ = Watch out! Common mistake
- ♻ = Quick Troubleshoot
- 📷 = Camera tip
- ✗ = Settings tutorial
- 🔒 = Privacy/Security alert

QUICK START PULLOUT SECTION

A 5-Minute Crash Course to Start Using Your iPhone 16 Right Away

If you're eager to jump in fast, here's your rapid-fire walkthrough to get up and running:

🔌 1. Turn It On & Set Up

- Press and hold the right-side button to power on

- Select your language, region, and connect to Wi-Fi

- Sign in or create an Apple ID

- Set up Face ID or Passcode

- Choose your preferences for Siri, iCloud, and Location Services

🏠 2. Understand the Home Screen

- Swipe up to go home

- Swipe down from the top-right for Control Center

- Tap and hold apps to rearrange or delete

- Swipe left or right to move between app pages

📞 3. Make a Call

- Open the **Phone app** → Tap **Keypad** → Dial a number → Press **Call**

💬 4. Send a Message

- Open **Messages app** → Tap the pencil icon ✏️

- Type a contact or number → Write your message → Tap **Send**

📷 5. Take a Photo

- Open **Camera app** → Point and tap the **white shutter button**

🔋 6. Charge Your iPhone

- Plug **the USB-C cable** into your iPhone
- Use a certified adapter or **MagSafe** charger for fast charging

🔐 7. Keep It Secure

- Go to **Settings** → **Face ID & Passcode**
- Set a secure passcode
- Enable Find My iPhone under your Apple ID for protection

This section is your quick Launchpad but everything is broken down in more depth later in the book, so come back anytime you need extra help, confidence, or pro tricks.

CHAPTER 1

Getting Started with iPhone 16

Unboxing iPhone 16: What's in the Box?

There's nothing quite like unboxing a new iPhone—the crisp pull of the tab, the sleek device tucked neatly in its tray, the fresh scent of untouched tech. Whether it's your first time or your fifteenth, Apple's signature unboxing experience still has that wow factor.

Here's what you'll typically find inside your iPhone 16 box:

- iPhone 16 itself (wrapped in a thin protective film)
- USB-C to USB-C Cable (Apple's move to universal charging)

- Documentation (Quick Start guide, SIM ejector pin if applicable, and warranty details)
- No charger brick or earbuds included—Apple now sells them separately for environmental reasons

Tip for Beginners: Use your existing USB-C wall adapter from any recent iPhone or iPad—or purchase Apple's 20W power adapter for fast charging.

Everything included inside your new iPhone 16 box— featuring the iPhone 16, USB-C to USB-C charging cable, SIM ejector tool, and standard documentation.

Overview of Physical Features

Before we dive into setting things up, let's get to know your new device. The iPhone 16's physical layout is beautifully minimal, but every button has a purpose.

Here's what you'll find:

Right Side

- **Side Button (Power)** – Press to wake, hold to turn off, double-tap for Apple Pay

Left Side

- **Volume Up / Down** – Adjust ringer, media, and call volume
- **Ring/Silent Switch** – Toggle between sound and vibrate (now customizable in Settings for iPhone 16 Pro models)

Bottom

- **USB-C Port** – For charging, data transfer, and accessories
- **Speaker Grille** – For media and call audio

Back

- **Dual or Triple Camera Array** (depending on model)
- **Flash** – Useful for photos and notifications

Front

- **Dynamic Island** – Apple's redesigned pill-shaped area for real-time alerts, music, Face ID, and background tasks

For Seniors: Feel each button with your fingers before turning on the device—it helps get familiar with orientation and layout.

12MP FRONT CAMERA

RING/SILENT SWITCH

VOLUME BUTTONS

48MP MAIN CAMERA

FLASH

12MP ULTRA WIDE CAMERA

SIDE BUTTON

SPEAKER MICROPHONE

USB-C CONNECTOR

SPEAKER / MICROPHONE

Setting Up Your iPhone 16 for the First Time

Let's power up and get you fully set up, step by step. Don't worry, this part is designed to be simple, even if you've never touched an iPhone before.

Step 1: Turn It On

- Press and hold the **Side Button** until you see the Apple logo
- Wait for the **"Hello" screen** to appear

Step 2: Choose Language & Region

- Tap your preferred **language**
- Select your **country or region**

This customizes your date/time format, keyboard, and Siri preferences

Step 3: Connect to Wi-Fi

- Select your home Wi-Fi network
- Enter the Wi-Fi password

Your iPhone will use Wi-Fi to complete setup, download settings, and activate features.

Step 4: Set Up Your Apple ID

- If you already have one, **sign in**

- If not, tap **"Don't have an Apple ID?"** → Create one

This ID lets you download apps, access iCloud, sync photos, and use FaceTime or iMessage

Step 5: Set Up Face ID & Passcode

- Follow the animated instructions to scan your face
- Choose a secure **six-digit passcode**

Pro Tip: Use Face ID for fast unlocking, payments, and app security

What you'll see during your iPhone 16 setup

26

For Seniors: Enabling Accessibility During Setup

Apple has built the iPhone 16 with accessibility in mind. You don't have to be tech-savvy or young to enjoy this device, it can adapt to your needs.

During setup, tap **Accessibility Options** (bottom-right on the setup screen) to turn on features like:

- **Larger Text** – Make everything easier to read
- **VoiceOver** – Your iPhone speaks the items on screen
- **Touch Accommodations** – Customize how you interact with the screen
- **Spoken Content** – Hear what you type or select
- **Color Filters / Bold Text** – Improve clarity for low-vision users

Senior Tip: You can turn these on any time later too. Just *go to **Settings** → **Accessibility**.*

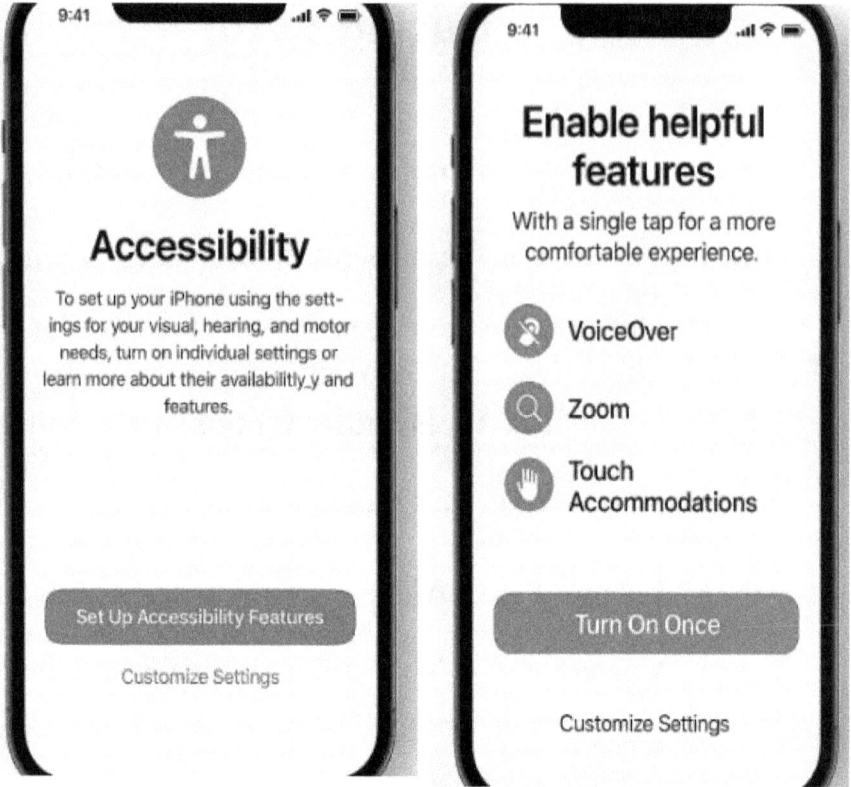

Chapter Wrap-Up Checklist

✓ Before we move on, let's confirm:

✓ You know what's inside the iPhone 16 box

✓ You've identified every button and port on your device

✓ You completed the initial setup (Wi-Fi, Apple ID, Face ID)

✓ Accessibility features are turned on if needed

You're now ready to explore the Home Screen and begin using your iPhone for calls, texts, photos, apps, and more. In the next chapter, we'll walk through the most important new iOS 18 features—and how to make the iPhone 16 *feel like home.*

CHAPTER 2

Understanding iOS 18—What's New & What Matters

If iPhone 16 is the body, **iOS 18** is the brain.

Apple's newest operating system isn't just a refresh, it's a leap toward greater personalization, smoother multitasking, and smarter automation. Whether you're a curious first-timer or a seasoned iOS user, this chapter will show you exactly what's new and why it matters.

Let's walk through the key features that define iOS 18 and how they can make your iPhone work harder for you.

iOS 18 Key Features Overview

Here's a quick snapshot of what's fresh and powerful in iOS 18:

- **A redesigned Control Center** with custom layouts and deeper app control
- **Interactive widgets** that you can use directly from your Home or Lock screen
- **Smarter Lock Screen customization** (fonts, colors, widgets, even images)
- **StandBy Mode**, which turns your iPhone into a smart display while charging
- **AI-powered suggestions and text predictions** woven across apps
- **Smarter Messages** with scheduled sends, message editing, and emoji reactions
- **Improved battery optimization** and app usage tracking
- **Locked & Hidden apps** for sensitive content
- **App Library enhancements** for faster app access

Throughout this chapter, we'll break these down in real-world terms so you can see the *difference and use it daily.*

31

Control Center Redesign

iOS 18 brings the **biggest change to Control Center in years**. Think of it as your iPhone's command panel—where you toggle Wi-Fi, flashlight, Bluetooth, volume, and more.

What's New:

- **Customizable Layout**: Long-press any toggle and tap "Customize Controls"
- **Multiple Pages**: Swipe between **custom control panels**—for Home, Music, Work, etc.
- **New Controls**: Add Notes, Shazam, Wallet, Sleep Mode, and even third-party apps
- **Resizable Controls**: Make your most-used toggles bigger or smaller

Every user now has a personalized Control Center. You can build your own layout to suit how you use your iPhone.

Lock Screen Customization

Apple now lets you **customize the Lock Screen** like never

before, blurring the line between function and style.

What You Can Do:

- Change **font style and color** of date & time

- Add **interactive widgets** (calendar, weather, reminders, photos, battery)

- Customize **multiple Lock Screens** and switch between them with a swipe

- Link **Focus modes** to specific Lock Screens (e.g., a calm lock screen during work)

Example: Create a "Work Lock Screen" that shows your calendar and emails, and a "Weekend Lock Screen" with music and fitness widgets.

StandBy Mode Explained

StandBy Mode is one of the most useful hidden gems in iOS 18 and it activates automatically when your iPhone is:

- Plugged into a charger

- In landscape (horizontal) position

- Locked

This mode transforms your iPhone into a **smart display**, perfect for:

- A digital clock (analog or full-screen options)

- A rotating photo frame

- Siri suggestions and calendar previews

- Real-time widgets (weather, reminders, etc.)

Great for your nightstand, desk, or kitchen counter—StandBy makes your iPhone useful even when idle.

Senior Tip: Enlarge the clock for better visibility while charging overnight.

iPhone in StandBy mode on a nightstand

Widgets and Smart Stack

Widgets are no longer just previews, they're **fully interactive** in iOS 18.

You can now:

- **Check off tasks** in Reminders without opening the app
- **Play music** directly from the widget
- **Scroll through Smart Stack** with a swipe
- Add widgets to the **Lock Screen** and **StandBy view**

The **Smart Stack** uses AI to show you the right widget at the right time:

- Calendar in the morning
- Weather when rain is near
- Maps when you have a trip planned
- Battery % when AirPods are connected

Pro Tip: Long-press on a widget → Choose "Edit Stack" → Rearrange or remove cards based on your routine.

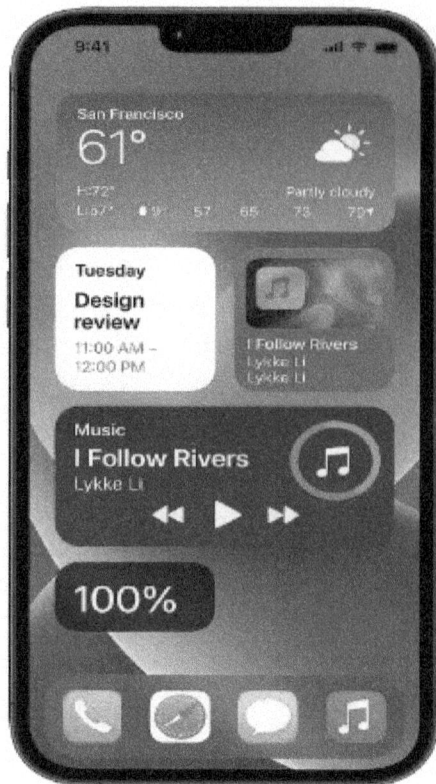

PRO TIP: Customize iOS for Productivity or Minimalism

Whether you're a multitasker or a digital minimalist, iOS 18 has tools to tailor your device to your lifestyle.

For Productivity:

- Set up **Focus Modes** with custom Lock Screens & app filters
- Create a **Work Control Center Page** (with Do Not Disturb, Calendar, Notes, Slack, etc.)
- Use **Shortcuts app** to automate routines (like "Start Work" or "Leave Office")

For Minimalism:

- Use the App Library to **hide unused apps**
- Limit screen time and alerts with **Focus Filters**
- Customize the Home Screen with **only essential widgets**
- Turn on grayscale or dim display during night hours

The key with iOS 18 isn't doing more—it's doing what *matters most to you*—better, faster, and with less friction.

Chapter Recap

You've just discovered:

- ✓ The smartest, most customizable Control Center yet

✓ How to create Lock Screens that match your day or mood

✓ How StandBy Mode turns idle time into useful time

✓ The power of interactive widgets and Smart Stack AI

✓ Ways to tailor iOS 18 to boost either focus or simplicity

In the next chapter, we'll shift from settings to skills—starting with iPhone 16 Basics for Beginners, where we guide you through everyday actions like gestures, apps, typing, and notifications.

CHAPTER 3

iPhone Basics for Dummies & Beginners

Whether this is your very first iPhone or you've just never had time to explore it, this chapter will make everything feel simple. No guesswork. No confusion. Just **tap**, **swipe**, and **go**.

Let's break down your iPhone's essential functions in a way that's clear, easy to follow, and empowering—no matter your age or experience.

Navigating with Gestures

With iPhone 16, Apple has fully committed to a buttonless interface—so understanding gestures is the key to mastering your device.

Here's how to move around your iPhone like a pro:

Go Home

- Swipe **up** from the bottom of the screen.

Switch Between Apps

- Swipe **left or right** along the bottom edge.

View Open Apps

- Swipe **up and pause** to reveal the App Switcher.

Access Notifications

- Swipe **down from the top-left**.

Open Control Center

- Swipe down from the top-right.

Senior Tip: Practice slow swipes—iPhones respond better to fluid movements, not fast flicks.

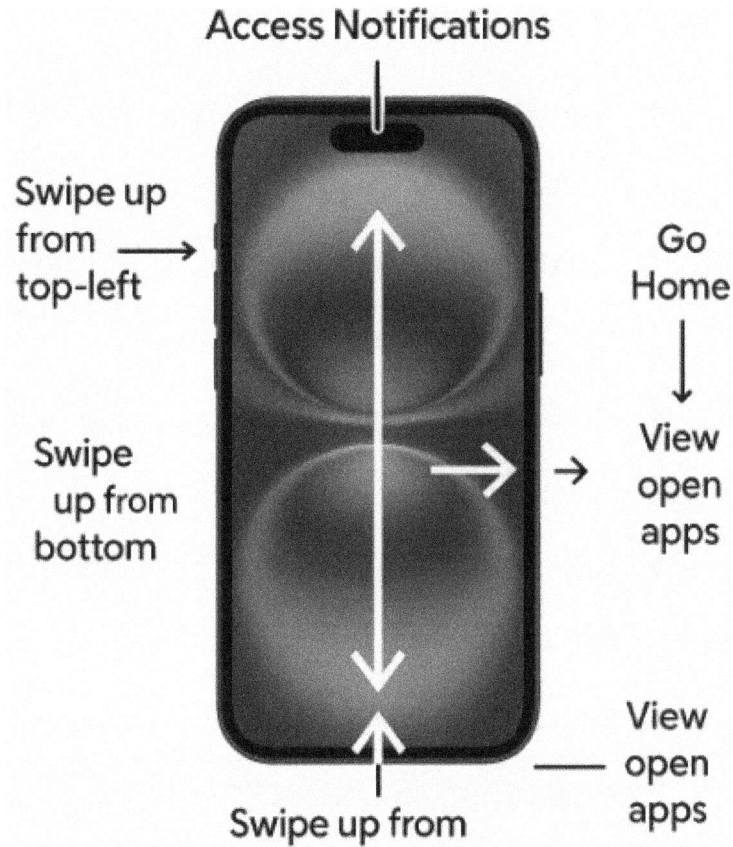

Access Notifications

Swipe up from top-left →

Go Home

Swipe up from bottom

View open apps

View open apps

Swipe up from

Understanding the Home Screen

Layout

Your **Home Screen** is your command center where all your apps live.

Here's what you'll find:

- **App Icons** – Tap to open, press and hold to move or delete

- **Smart Stack** – A rotating widget preview of apps like Weather, Calendar, Music

- **Dock (Bottom Bar)** – Fixed apps for quick access (Phone, Safari, Messages, Music)

- **Page Dots** – Indicates which Home screen page you're on

- **Search Bar (Swipe Down)** – Instantly find apps, contacts, or information

You can customize your Home Screen with folders, widgets, or minimal layouts to suit your lifestyle.

Opening and Closing Apps

To Open:

- Simply **tap any** app icon.

To Close:

- Swipe **up and pause**, then **swipe the app** away (upward) from the App Switcher.

You don't need to close every app constantly—iPhones manage memory automatically.

Create a Folder:

- Drag one app **on top of another**

- It creates a folder—name it something like "Work" or "Games"

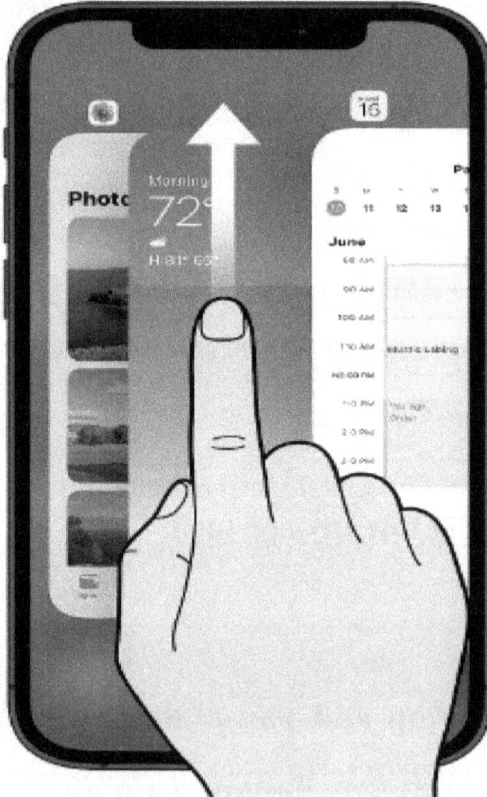

Using the Keyboard (Including Dictation)

Typing on an iPhone might feel tricky at first but it gets easier with time. And for those who don't want to type at all, **dictation** can help.

To Type:

- Tap inside any text box
- Use the on-screen keyboard that appears
- Tap **Shift (↑)** to capitalize
- Tap **Emoji (☺)** to add emoticons
- Tap **123** to access numbers/symbols

To Dictate:

- Tap the **Microphone** icon on the bottom-right of your keyboard
- Speak clearly. Example: "Hello comma I'm learning to use my iPhone period."

- Tap "Done" when finished

Voice dictation is helpful for those with limited dexterity or vision.

Start to speak when this symbol appears.

Tap to turn Dictation on and off.

Notifications, Control Center &

Spotlight Search

Notifications

- Swipe down from **top-left corner**

- See missed calls, texts, calendar alerts, reminders, etc.

- Swipe left on a notification to clear it, or press to respond

Control Center

- Swipe down from **top-right corner**

- Toggle Wi-Fi, Bluetooth, Do Not Disturb, brightness, flashlight, and more

- Tap and hold icons for more settings

Spotlight Search

- Swipe **down on Home Screen** (not from the top)

- Search your iPhone for apps, contacts, photos, websites, or answers from Siri

Notification Center

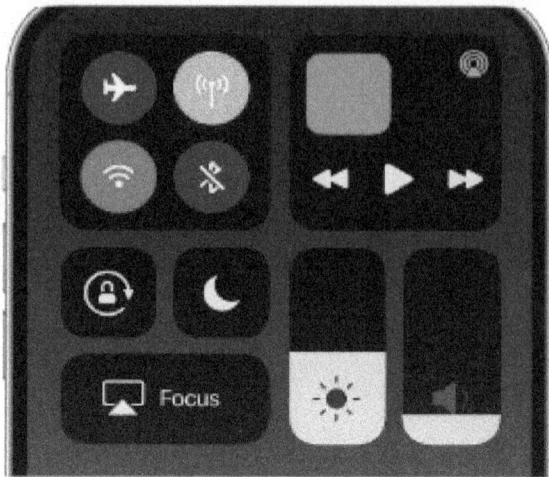

Control Center

For Seniors: Making Text Bigger &

Enabling Voice Feedback

Apple built powerful accessibility features that make using

an iPhone easier, clearer, and more comfortable.

Make Text Bigger:

1. Go to **Settings**

2. Tap **Accessibility** → **Display & Text Size**

3. Tap **Larger Text**

4. Slide the bar to increase size

Enable Voice Feedback:

1. Go to **Settings** → **Accessibility**

2. Tap **Spoken Content**

3. Turn on:

   - **Speak Selection** (tap text and hear it spoken)

   - **Speak Screen** (swipe down with two fingers to read full screen)

You can even adjust the voice's speed or accent to your liking.

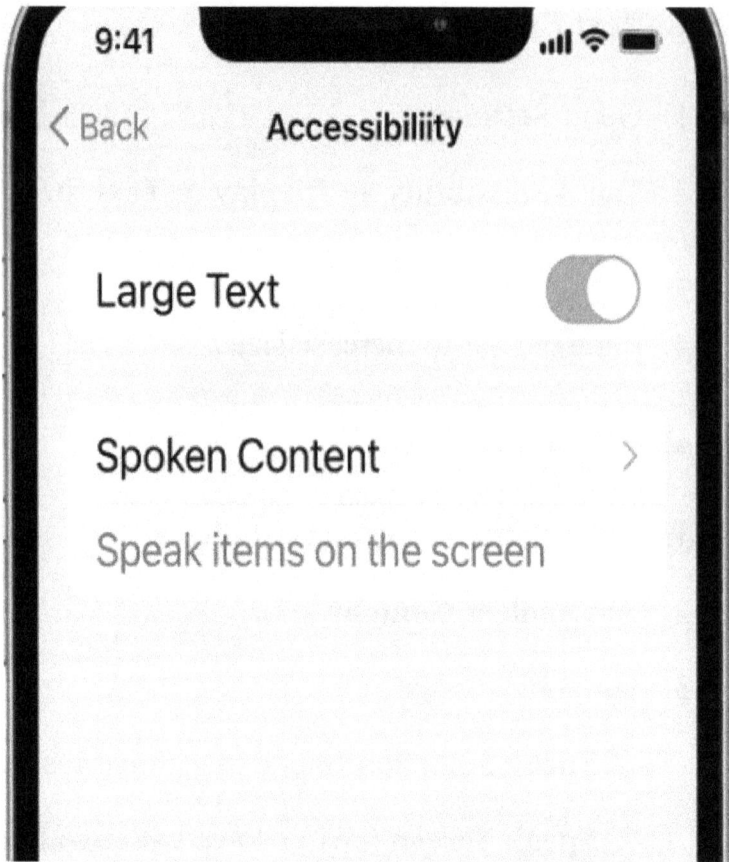

Chapter Summary

By now, you should feel confident doing the following:

✓ Swipe to navigate between apps, Control Center,
and more

✓ Understand your Home Screen and rearrange apps/folders

✓ Open, close, and organize apps easily

✓ Use the keyboard—or speak instead

✓ Check notifications and search for anything

✓ Make your screen easier to see or have text read out loud

Next, we'll explore Chapter 4: Making Calls, Sending Texts, and FaceTime—where communication becomes easy, fun, and intuitive.

CHAPTER 4

Making Calls, Sending Texts, and FaceTime

The heart of any smartphone especially an iPhone is staying connected with the people you care about.

In this chapter, you'll learn how to:

- Make phone calls with confidence

- Save and find contacts easily

- Send and receive text messages and iMessages

- Use FaceTime to video chat with friends or family

- Express yourself with emojis, photos, and group messages

Let's make your iPhone 16 a communication powerhouse with **no confusion or frustration**.

Adding and Saving Contacts

Storing your contacts in the **Phone** or **Contacts** app means you never have to remember phone numbers again.

To Add a New Contact:

1. Open the **Phone** app

2. Tap **Contacts** at the bottom

3. Tap the + sign (top right)

4. Enter the name, phone number, and email (optional)

5. Tap **Done**

You can add a photo to each contact to quickly recognize who's calling.

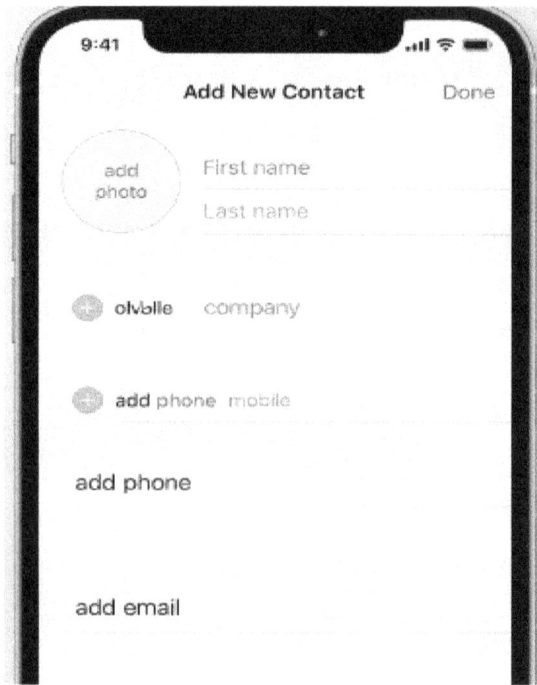

Dialing and Receiving Calls

There are three main ways to place a phone call:

From Contacts:

- Open the **Phone** app → Tap **Contacts** → Choose a contact → Tap **Call icon**

Using the Keypad:

- Open **Phone** → Tap **Keypad** → Dial the number → Tap **Green Phone icon**

From Recents:

- Tap **Recents** tab in Phone app → Tap the number or name

To Answer a Call:

- If the phone is **unlocked**, tap **Green button** to answer
- If **locked**, swipe the **slide to answer** slider

Senior Tip: Set your phone to announce callers under Settings → Phone → Announce Calls.

Sending SMS and iMessage

The **Messages** app lets you send simple texts, photos, videos, emojis, and more.

Here's the difference:

- **SMS** = Regular text message (green bubbles)

- **iMessage** = Apple-to-Apple messages (blue bubbles), requires Wi-Fi or cellular data

To Send a Message:

1. Open the **Messages** app
2. Tap the **Pencil icon** in the top right
3. Enter a phone number or contact name
4. Type your message in the box
5. Tap the **Blue arrow** to send

iMessages are free when sent over Wi-Fi—great for international texting.

Using FaceTime (Audio & Video)

FaceTime is Apple's free video or audio calling service. You can see and talk to loved ones anywhere—perfect for grandparents, travelers, or just saying hi.

Start a FaceTime Call:

1. Open the **FaceTime app**

2. Tap **New FaceTime**

3. Enter a name, number, or email

4. Tap either:

   - **Video icon** for a video call

   - **Phone icon** for audio-only FaceTime

FaceTime works over Wi-Fi or mobile data—no minutes required.

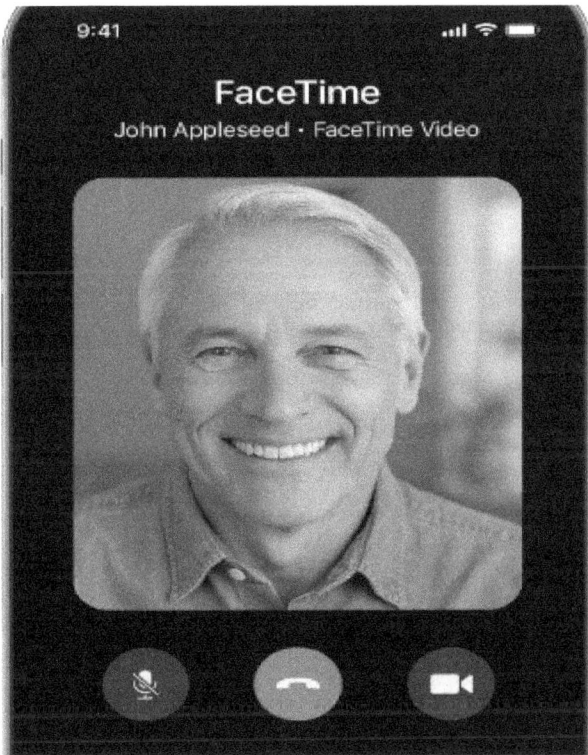

Group Messaging & Emojis

Want to chat with the whole family? Or drop a birthday message filled with 🎉 emojis?

Start a Group Text:

1. Open **Messages**

2. Tap the **Pencil icon**

3. Add **multiple contacts** in the "To" field

4. Send your message

Group messages work better when everyone uses iPhones (iMessage)

Using Emojis:

- Tap the Smiley face icon on the keyboard

- Choose from hundreds of expressions, animals, objects, and symbols

- Tap to insert, then press **Send**

Pro Tip: Type something like "love" or "birthday" and emoji suggestions will appear above your keyboard.

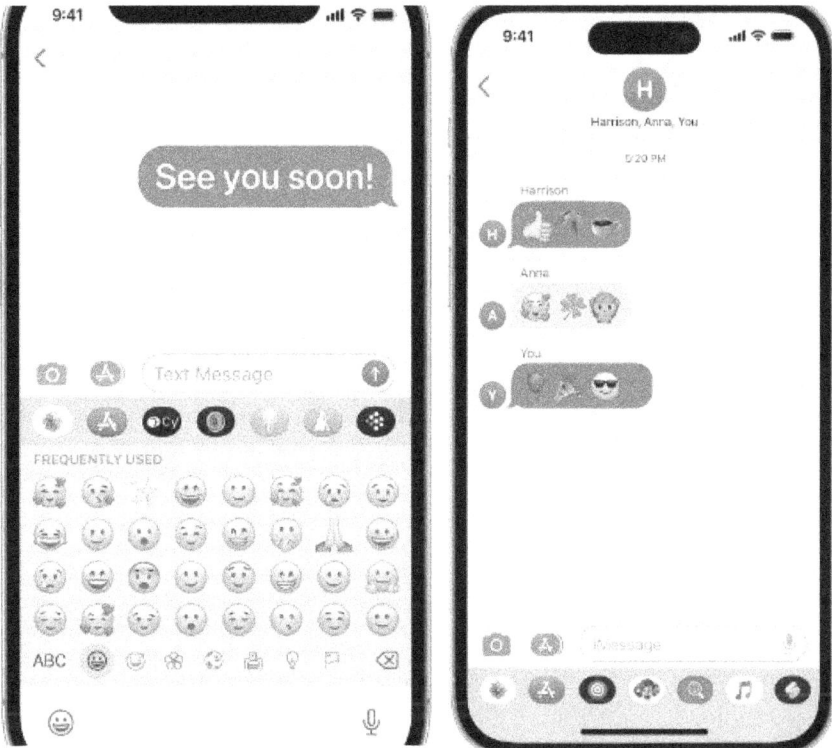

Chapter Summary

By now, you've learned how to:

- ✓ Add and manage your contact list
- ✓ Make and receive phone calls effortlessly
- ✓ Send texts, iMessages, and group messages

✓ Use FaceTime to video call loved ones

✓ Add fun, meaningful emojis to express yourself

In the next chapter, we'll explore the iPhone 16 Camera—and how to take stunning photos and videos with ease.

CHAPTER 5

Mastering the iPhone 16 Camera

The camera on the iPhone 16 isn't just a lens—it's a tool that lets you capture memories, moments, moods, and masterpieces.

Whether you're shooting birthday portraits, vacation landscapes, videos for social media, or simple everyday snapshots, this chapter will help you unlock every essential feature from basic functions to professional tools—step by step.

Opening the Camera App: Portrait, Video, Slo-Mo & Cinematic

Let's start by getting comfortable with the **Camera app interface**.

How to Open the Camera:

- Tap the **Camera app** icon on your Home Screen

- OR swipe left on your Lock Screen

- OR swipe down for **Control Center**, then tap the **Camera icon**

Once open, swipe left or right at the bottom to switch between camera modes:

Photo (Default):

- For everyday pictures—tap once to focus, press shutter button.

Portrait Mode:

Adds **depth effect (blurred background)** like a DSLR. Great for people, pets, or objects.

- Use **Natural** Light, Studio Light, Contour Light, etc.

Video Mode:

Record video in HD or 4K.

- Press red **record button** to begin

- Tap anywhere to focus or swipe up for exposure

Slo-Mo:

- Record **slow-motion** videos, great for action shots or creative effects.

Cinematic Mode:

This is iPhone 16's next-level feature for video storytelling.

- Automatically shifts focus from foreground to background

- You can adjust focus **even after recording**

Use Cinematic for interviews, B-roll, or TikToks with that cinematic vibe.

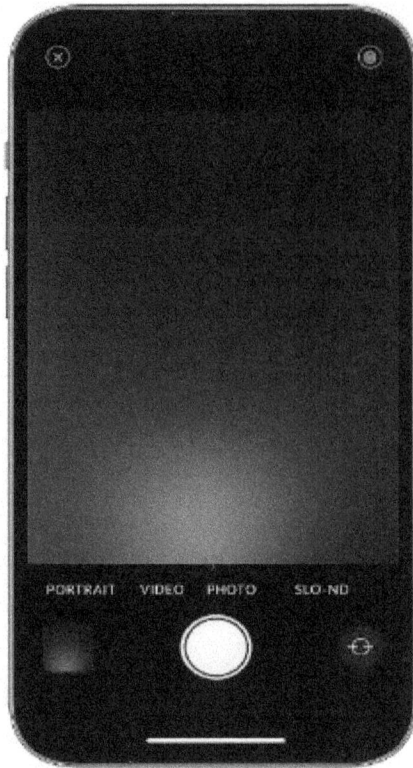

How to Use Photographic Styles, Zoom & Night Mode

Apple gives you powerful tools to **customize how your photos look**—before you even snap them.

Photographic Styles:

Go to **Settings** → **Camera** → **Photographic Styles**

- Choose from: **Standard, Rich Contrast, Vibrant, Warm, or Cool**
- These are NOT filters, they permanently change the image tone

You can even adjust **tone and warmth** with sliders before taking the shot.

Zoom Tips:

- Use **Pinch to Zoom** gesture or tap the **.5x, 1x, 2x, 5x** buttons
- iPhone 16 Pro models support **optical zoom**, giving sharper detail without losing quality

Night Mode:

- Automatically activates in low light
- Yellow moon icon appears when active
- Hold steady—Night Mode uses long exposure for better brightness and clarity

Try using a tripod for best night shots (stars, city lights,

candlelight)

Editing Photos (Crop, Filters, Markup)

After you take a photo, the magic continues with Apple's **built-in editing tools**.

To Edit a Photo:

1. Open the **Photos app**

2. Tap the photo

3. Tap **Edit** (top-right)

Now you can:

- Crop & straighten

- Adjust brightness, contrast, shadows, and highlights

- Apply built-in filters

- Tap **Markup** to draw, highlight, add text or shapes

Undo anytime by tapping "Revert" to go back to the original.

Using AI Tools in the Photos App

(Subject Lift, Live Text)

The iPhone 16 includes powerful AI features that turn your photos into **smart, searchable content**.

Subject Lift (Remove Background):

1. Tap and hold on a **person or object** in a photo

2. The subject will **lift from the background**

3. Drag to other apps (like Messages, Notes, or Mail)

Live Text:

iPhone recognizes text in your photos or camera view.

Use it to:

- Copy text from signs, books, screenshots

- Translate foreign languages

- Dial phone numbers directly from business cards

Try pointing your camera at a menu or a street sign—it will highlight the text automatically.

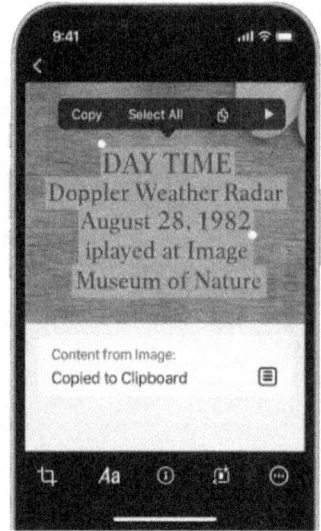

Organizing Albums

Photos can pile up fast. Here's how to **stay organized**:

Create an Album:

1. Go to **Photos app** → **Albums**

2. Tap + in the top-left corner

3. Choose **New Album**

4. Name it (e.g., "Holidays," "Family," "Dog Pics")

5. Add photos from your library

Tips for Managing Albums:

- Use **Favorites** (♥□) to quickly find your best shots

- **People & Places** albums are created automatically

- Search by **date, keyword, or object** ("beach", "cake", "car")

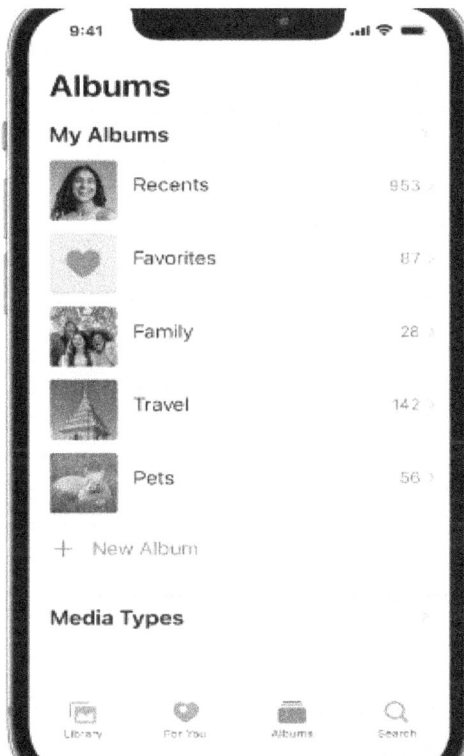

PRO USER TIP: Master Composition

for Stunning iPhone Photos

You don't need to be a professional to take **professional-looking photos**—you just need to understand the basics of *composition*:

Rule of Thirds:

- Turn on the **Grid** in Settings → Camera
- Place your subject along the lines or intersections—not in the center

Leading Lines:

- Use roads, fences, shadows, or hallways to **draw the eye** into your subject

Fill the Frame:

- Get closer to avoid distractions and add emotional impact

Declutter the Background:

- Keep it simple—your subject should pop

Light Is Everything:

- Face your subject toward the light (natural light works best)
- Avoid harsh shadows or backlighting (unless it's intentional)

Great composition turns a quick photo into a stunning image—no expensive gear required.

Chapter Summary

Here's what you now know how to do:

- ✓ Navigate and use the camera's major shooting modes
- ✓ Customize your shots with styles, zoom, and Night Mode
- ✓ Edit photos like a pro with filters, crop, and markup
- ✓ Use subject lift and Live Text to unlock AI features
- ✓ Stay organized with albums and searchable tags

✓ Apply photography composition rules for gallery-worthy photos

Next up in Chapter 6, we'll explore Siri and Shortcuts—the automation tools that make your iPhone not just smart, but brilliantly helpful.

CHAPTER 6

Siri & Shortcuts – Automate Your iPhone

Imagine your iPhone doing the little things for you without you lifting a finger. That's the magic of **Siri** and **Shortcuts**.

From sending messages and setting reminders to starting your morning playlist or controlling your smart home, Siri is more than a voice assistant, it's **your hands-free digital helper**, built into your iPhone 16.

In this chapter, we'll show you how to:

- Set up Siri and customize it to your liking
- Use Siri for everyday tasks like calling, texting, and creating notes
- Create *your own* Siri-powered automations with Shortcuts
- Go pro with **smart home integration**

Setting Up and Talking to Siri

If you've never used Siri before, don't worry—setup only takes a minute.

How to Set Up Siri:

1. Open **Settings**

2. Tap **Siri & Search**

3. Turn on:

   - **Listen for "Hey Siri"**

   - **Press Side Button for Siri**

   - **Allow Siri When Locked**

4. Follow the on-screen prompts to train Siri with your voice

You can now activate Siri by:

   - Saying **"Hey Siri"**

   - Holding the **side button**

Senior Tip: If saying "Hey Siri" feels awkward, you can just press and hold the side button instead.

What You Can Say:

Siri understands **natural speech**—just speak like you're talking to a person.

Try these:

- "Call Anna."
- "Send a message to Dad saying I'll be late."
- "What's the weather today?"
- "Remind me to take my medicine at 7 PM."
- "Turn on the flashlight."
- "Set a timer for 10 minutes."

Daily Uses: Calls, Texts, Reminders, App Launches

Siri saves you time by handling frequent tasks without you

needing to navigate menus.

📞 Make Calls Hands-Free:

- "Hey Siri, call John mobile."
- "Redial the last number."

💬 Send Messages:

- "Text Sarah: I'll be there in 15 minutes."

🗓 Set Reminders:

- "Remind me to take a walk every evening at 6 PM."
- "Remind me when I get home to check the oven." (location-based reminders)

🚀 Launch Apps or Settings:

- "Open Camera."
- "Open WhatsApp."
- "Turn on Low Power Mode."

Location-Based Examples:

- "Remind me to pick up milk when I leave work."

- "Text my wife when I arrive at the gym."

Use Siri while driving, cooking, walking, or relaxing—you'll realize how much easier life becomes.

Creating Custom Siri Shortcuts (With Real-World Examples)

Want Siri to do **a sequence of tasks with one command**? That's where Shortcuts comes in.

Think of a shortcut as a **pre-programmed routine**—you tap a button or say a phrase, and your iPhone does multiple things automatically.

To Create a Shortcut:

1. Open the **Shortcuts** app (built into iOS)
2. Tap + to create a new shortcut
3. Tap **Add Action**
4. Choose actions like:

- Open app

- Send message

- Play music

- Get directions

5. Give it a name (e.g., "Morning Routine")

Real-Life Shortcut Ideas:

Morning Routine

Say: "Hey Siri, start my day"

- iPhone does:

- Reads weather

- Turns off Do Not Disturb

- Plays music or podcast

- Opens Calendar

Driving Mode

Say: "Hey Siri, let's drive"

iPhone does:

- Opens Maps

- Sends ETA to spouse

- Starts your driving playlist

Grocery Trip

Tap shortcut

iPhone does:

- Opens Notes (with your grocery list)

- Turns on Low Power Mode

- Opens your favorite shopping app

Good Night Shortcut

- Turns off Wi-Fi

- Sets alarm for 7 AM

- Enables Do Not Disturb

- Dims screen brightness

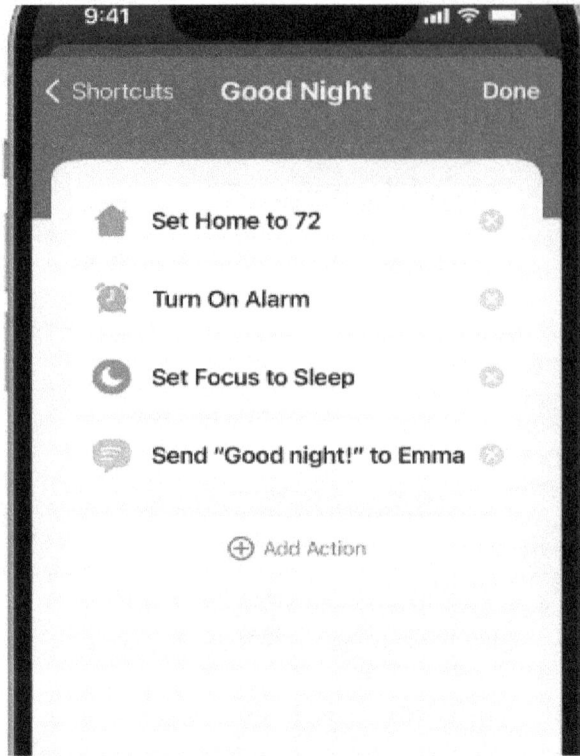

You can add your shortcuts to the Home Screen or trigger them with Siri voice commands.

Pro Corner: Use Siri with HomeKit and Smart Home Integration

If you have smart home devices like lights, plugs,

thermostats, or locks, **Siri becomes your home's voice-activated control panel**.

Siri + HomeKit Examples:

- "Hey Siri, turn on the kitchen lights."
- "Hey Siri, lock the front door."
- "Set the thermostat to 72 degrees."
- "Open the blinds."
- "Turn off everything" (bedtime mode)

To Set Up Home Automation:

- Open the **Home app**
- Tap + to add a device (must be HomeKit compatible)
- Create **Scenes** or **Automations** like:
  - "Movie Time" – dims lights, lowers blinds, plays TV
  - "Arrive Home" – unlocks door, lights up porch

Ask Siri to run your scenes, and your iPhone becomes a

remote control for your whole house.

Chapter Recap

You now know how to:

- ✓ Set up and talk to Siri naturally
- ✓ Use Siri for daily tasks like messaging, calling, reminders, and launching apps
- ✓ Create custom Shortcuts for personal routines
- ✓ Control smart home devices hands-free with Siri + HomeKit

You've just unlocked the power of automation on iPhone 16 making your phone truly work for you.

Next, in Chapter 7, we'll explore Safari, Mail, Messages & Everyday Apps—so you can master the tools you use every day.

CHAPTER 7

Safari, Mail, Messages & Everyday Apps

Your iPhone 16 comes packed with powerful default apps that cover everything from web browsing to staying organized, chatting with family, and checking the weather. In this chapter, we'll help you make the most of these **everyday essentials**—even if you're brand new to them.

Browsing with Safari: Tabs, Reading List & Private Browsing

Safari is the iPhone's built-in web browser—your gateway to the internet.

To Open Safari:

- Tap the **Safari icon** (blue compass)

- Type a website or search keyword into the address bar

Managing Tabs:

- Tap the **Tab icon** (two squares, bottom right)
- Swipe left or right to view all open pages
- Tap the + **icon** to open a new tab
- Press and hold a tab to **reorder, close all**, or **group tabs**

Using Reading List:

- Found something useful but don't have time to read?
- Tap the **Share icon** (square with arrow)
- Select **Add to Reading List**
- Later, tap the **book icon** → Reading Glasses icon to view saved pages

Private Browsing Mode:

Tap **Tabs icon** → Tap "Private"

This mode doesn't save history, cookies, or autofill info

Use Private Mode for banking, gift shopping, or other sensitive browsing.

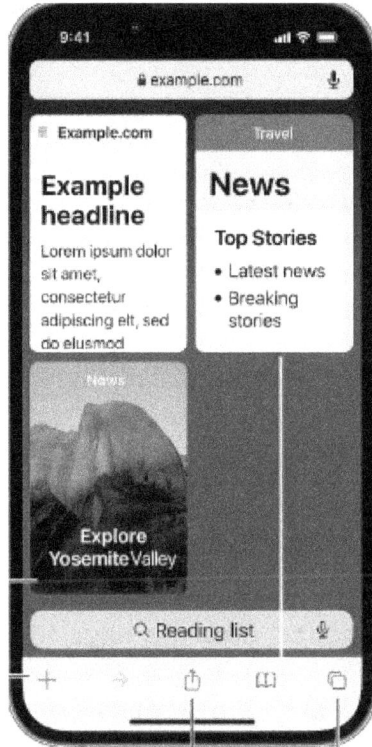

Setting Up Email Accounts

Mail on iPhone supports Gmail, Outlook, Yahoo, iCloud,

and custom emails.

To Set Up Mail:

1. Open **Settings** → **Mail** → **Accounts**

2. Tap **Add Account**

3. Choose provider (e.g., Gmail)

4. Enter your email and password

5. Toggle services you want to sync (Mail, Contacts, Calendars)

Your email will now be available via the **Mail app** on your Home Screen.

Tips for Using Mail:

- Swipe left on a message to archive or delete

- Tap **Flag icon** to mark important mail

- Pull down to refresh your inbox

- Tap the **filter icon** to show unread or flagged mail only

You can add multiple email accounts and switch between

inboxes.

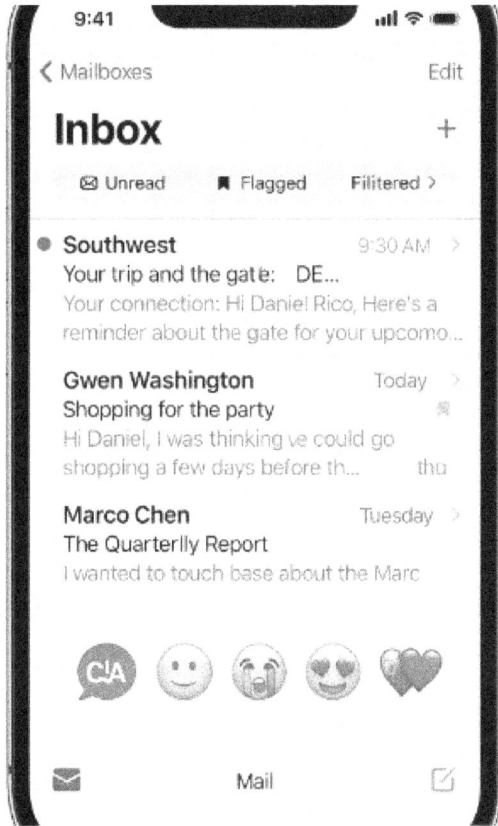

Tips for iMessage & Group Chats

You've already learned the basics of Messages, but let's take it a step further.

Share Content:

- Tap the **camera icon** to send a live photo or video

- Tap the **App Store icon** to send stickers, GIFs, music, or app content

Group Messages:

- To name a group: Tap group → Tap arrow > Info → Name field

- Mute noisy chats: Tap group → Info → **Toggle Hide Alerts**

- Share your location: Info → Tap Share My Location

Voice Messages:

- Tap the **Microphone** icon beside the typing field

- Speak and send voice snippets (great for quick replies!)

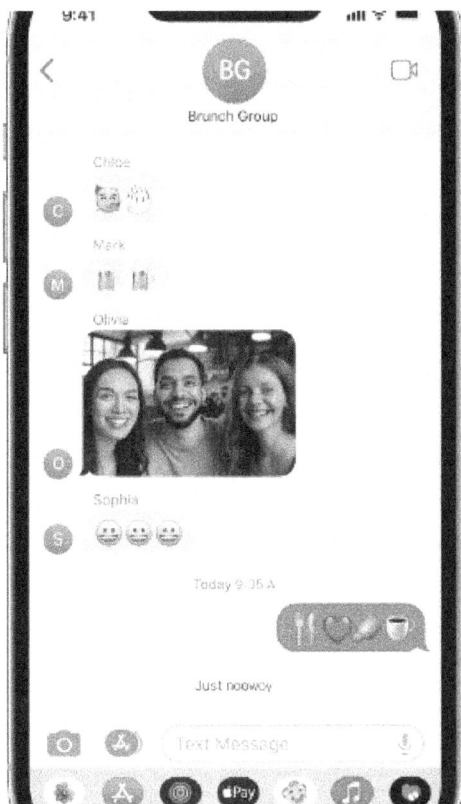

Using Notes, Reminders, Calendar

& Weather

These tools are simple, yet incredibly powerful when used right.

Notes:

Open the **Notes app**

- Tap **New Note** (▬·☐) icon
- Type, paste, or scan text
- Tap **camera icon** to scan documents
- Create folders to stay organized

Use Notes for shopping lists, ideas, journaling, or storing important info.

Reminders:

- Open the **Reminders app**
- Tap New **Reminder** (+)
- Add title, time, or location
- Set recurring reminders for tasks like medications or bills

Use Siri to create reminders hands-free.

Calendar:

- Open **Calendar app**

- Tap + to add events

- Set alerts, time zones, travel time

- Use different calendar colors for work, family, and personal

Weather:

- Open **Weather app**

- Tap cities to view forecasts

- Tap **Map icon** for real-time radar

- Add favorite cities by tapping **list icon** → +

The iOS 18 Weather app includes sunrise/sunset times, UV index, and air quality data.

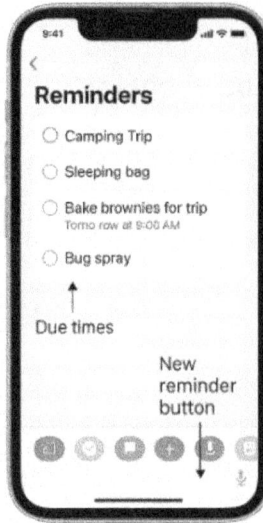
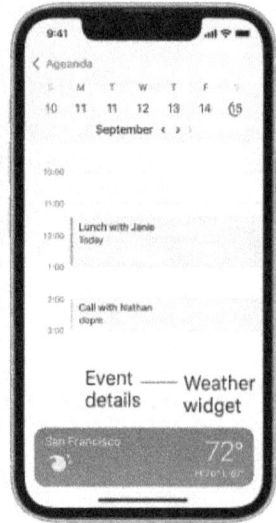

| Notes | Reminders | Calndar/Weather |

Chapter Summary

You now know how to:

- ✓ Browse the web using Safari's tabs, reading list, and private mode

- ✓ Set up and manage multiple email accounts with ease

- ✓ Make group messages and voice replies more fun and functional

- ✓ Stay organized using Notes, Reminders, Calendar, and Weather apps

These are the tools that help your iPhone 16 become a **daily assistant**—not just a communication device.

In Chapter 8, we'll dive into App Store, Widgets, and Productivity Essentials—showing you how to download apps, boost focus, and build a smarter Home Screen.

CHAPTER 8

App Store, Widgets, and Productivity Essentials

Your iPhone 16 isn't just a tool for communication, it's a gateway to **thousands of apps**, widgets, and smart features that help you organize your life, stay focused, and get more done each day.

Whether you want to check the news at a glance, organize your apps, or build a personal "Work Mode" that blocks distractions, this chapter will help you customize your iPhone into a **personal command center**.

Downloading & Managing Apps

The **App Store** is where you find apps for almost anything—shopping, meditation, games, banking, fitness, social media, recipes, you name it.

To Download an App:

1. Open the App Store

2. Tap the **Search icon** at the bottom

3. Type the name of the app you want (e.g., "Zoom" or "Facebook")

4. Tap **Get** (or the price if it's a paid app)

5. Authenticate with **Face ID** or your Apple ID password

To Delete an App:

- Press and hold the app icon

- Tap **Remove App**

- Choose **Delete App**

To Update Apps:

- Open App Store → Tap your **profile picture** (top-right)

- Scroll to see pending updates → Tap **Update All**

Most updates happen automatically, but it's good to check

for improvements.

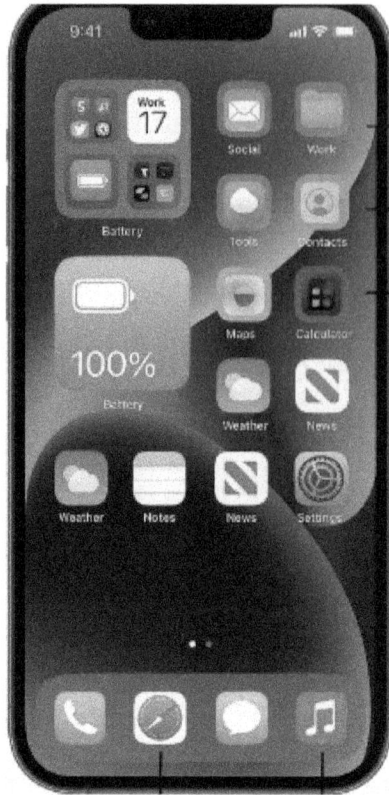

Organizing Apps into Folders

You don't have to keep swiping through pages of scattered apps. You can organize your apps into **folders** based on your life.

To Create a Folder:

1. Press and hold an app icon

2. Drag it on top of another app

3. A folder will appear—name it (e.g., "Finance," "Games," "Work")

4. Drag more apps into the folder

Keep your Dock clean with only 4-5 apps you use daily. Store everything else in folders.

Using Widgets for Weather, Tasks, Calendar & News

Widgets let you see info at a glance without opening apps. They live on your Home Screen or Today View (swipe right from Home).

To Add a Widget:

1. Press and hold a blank spot on the Home Screen

2. Tap the + **icon** in the top-left

3. Search or scroll to select a widget (e.g., Weather, Calendar, Reminders)

4. Tap **Add Widget,** then drag it into place

Popular Widgets:

- ▦ **Calendar**: See your next meeting or event

- ☁ **Weather**: Get live updates for your location

- ⬜ **Reminders**: See daily to-dos

- ▤ **News**: Read headlines without launching the app

- ▬ **Battery**: Check iPhone + AirPods charge at a glance

Senior Tip: Use large-size widgets with bold text for better visibility.

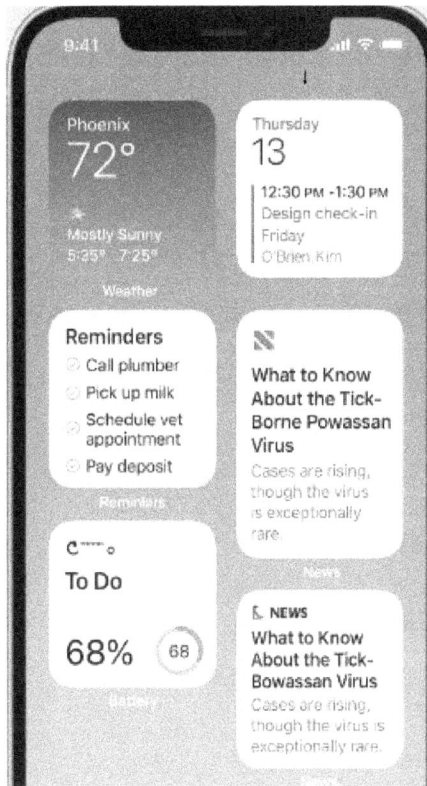

Screen Time & Focus Mode Setup

Sometimes, staying focused means **controlling distractions**. Apple gives you tools to **track usage** and **limit interruptions** with **Screen Time** and **Focus Modes**.

Screen Time:

Shows how much time you spend on apps, and allows you to set limits.

To Set Up:

1. Go to **Settings** → **Screen Time**
2. Turn it on → Choose "This is My iPhone"
3. Tap **App Limits** → Set daily limits for social media, games, etc.
4. Use **Downtime** to block all apps during rest or work hours

Focus Mode:

Helps you stay in control by silencing specific notifications and apps.

To Set Up:

1. Open **Settings** → **Focus**
2. Tap + to create a new mode (e.g., "Work" or "Sleep")
3. Choose:

- Allowed notifications (from people or apps)

- Home Screen settings (hide distractions)

- Lock Screen options (dim screen, silence alerts)

You can also **automate Focus Mode** to activate during:

- A certain time

- Location (e.g., office)

- App usage (e.g., open Zoom → enter "Work Mode")

Tip: Create a "Work" and "Home" Profile Using Focus

Work Mode:

- Silence social media and personal texts

- Show only work apps on Home Screen

- Auto-reply to calls: "I'm working. Will respond soon."

Home Mode:

- Silence email and work notifications

- Highlight family, music, photos, and relaxation apps

- Enable Do Not Disturb during dinner or bedtime

Use Smart Activation to switch profiles automatically based on time or location.

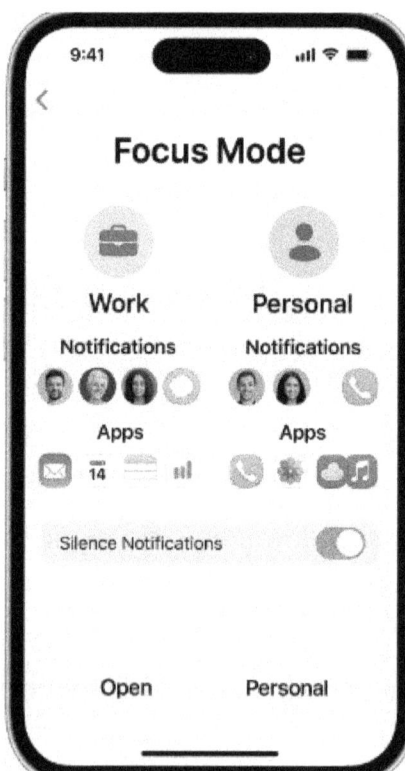

Chapter Summary

Now your iPhone can do more with less effort:

- ✓ Download, manage, and update apps
- ✓ Organize your Home Screen into useful folders
- ✓ Add widgets for weather, calendar, tasks, and more
- ✓ Set screen limits and reduce distractions
- ✓ Build custom Focus Modes that switch based on time, app, or location

Up next in Chapter 9, we'll explore Security, Privacy, and Parental Controls—so your iPhone not only works hard, but also protects your data and your family.

CHAPTER 9

Security, Privacy & Parental Controls

In today's world, **privacy and safety aren't optional they're essential**. Your iPhone 16 comes with powerful tools that protect your identity, location, kids, and data from unwanted access.

This chapter will walk you through everything you need to:

- Lock your phone securely
- Control who can access your camera, mic, and location
- Set up parental controls
- Hide private content
- And for advanced users—tap into Apple's built-in VPN features

Face ID, Passcode & Lockdown Mode

These are your **first line of defense**. A secure phone starts with proper lockscreen protection.

To Set Up Face ID and Passcode:

1. Go to **Settings → Face ID & Passcode**

2. Set a **6-digit passcode** (or custom alphanumeric)

3. Tap **Set Up Face ID**

4. Follow the on-screen instructions to scan your face

You can choose to use Face ID for:

- Unlocking your phone

- App Store purchases

- Password Autofill

- Secure apps like banking or health

What Is Lockdown Mode?

Lockdown Mode is for **maximum security**—ideal if you're concerned about digital threats or spyware.

To Enable:

- Go to **Settings** → **Privacy & Security** → **Lockdown Mode**
- Toggle **Turn On Lockdown Mode**

This disables certain features temporarily:

- Blocks unknown FaceTime calls
- Limits JavaScript in Safari
- Restricts USB access

Pro Tip: Use this when traveling, during sensitive work, or if your phone is lost.

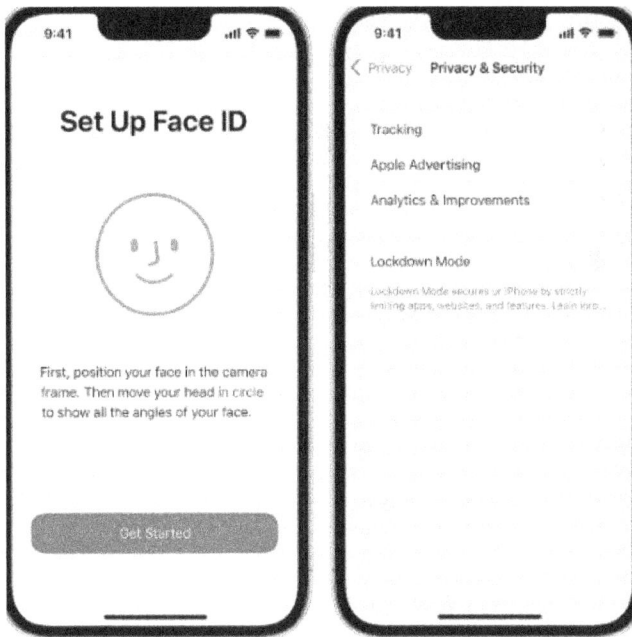

App Permissions: Location, Microphone, Camera

Many apps ask for access to your private data. You get to decide **who sees what**.

To Review or Change Permissions:

1. Go to **Settings → Privacy & Security**

2. Tap any category: **Location Services, Microphone, Camera, Contacts, etc**.

3. Toggle access for each app:

   - Never

   - Ask Next Time

   - While Using the App

   - Always (only when necessary)

Disable access for apps that don't truly need it—especially games, free tools, or untrusted developers.

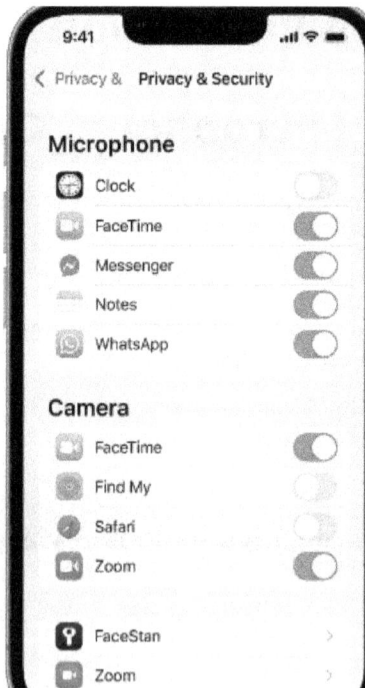

Screen Time for Kids: Set Limits & Restrictions

Apple's **Screen Time + Family Sharing** gives parents full control over what their children can access.

To Set Up Screen Time for a Child:

1. Go to **Settings → Screen Time**
2. Tap **Set Up Screen Time for Family**
3. Create or select your child's Apple ID
4. Set:
   - Downtime: When the device can't be used
   - App Limits: Set daily time caps (e.g., 1 hr for YouTube)
   - Content Restrictions: Block explicit content, web adult filters, app ratings

Content & Privacy Restrictions:

- Go to **Screen Time** → **Content & Privacy Restrictions**

- You can block:

  - Installing or deleting apps

  - In-app purchases

  - Access to adult websites

  - Siri web search

Senior Tip: Use this not just for kids, but to simplify the iPhone for elderly loved ones.

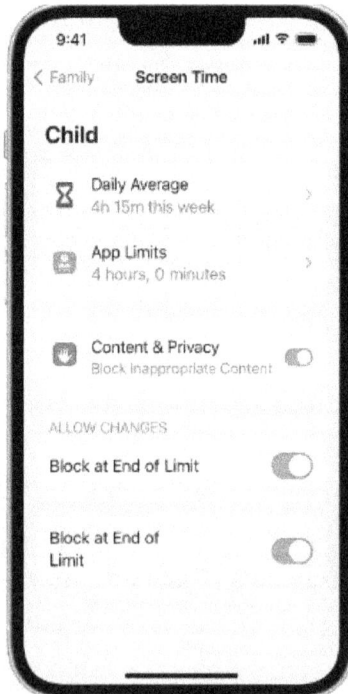

Hide Apps, Passwords & Private Content

iPhone 16 lets you **lock down private things** without needing third-party apps.

Hide Apps from the Home Screen:

- Long-press the app → Tap **Remove from Home Screen**
- It stays available in the **App Library** (swipe right → search)

Lock Notes with Password or Face ID:

1. Open **Notes** → **Select a note**
2. Tap **Share** → **Lock Note**
3. Use Face ID or a password

Hide Photos:

- Open **Photos** → **Select**

- Tap **Share → Hide**

- Find them in the **Hidden album** under **Photos →
  Utilities**

To lock Hidden or Recently Deleted albums:

- Go to **Settings → Photos → Use Face ID**

*All of this can be disabled or enabled later via your
Privacy settings.*

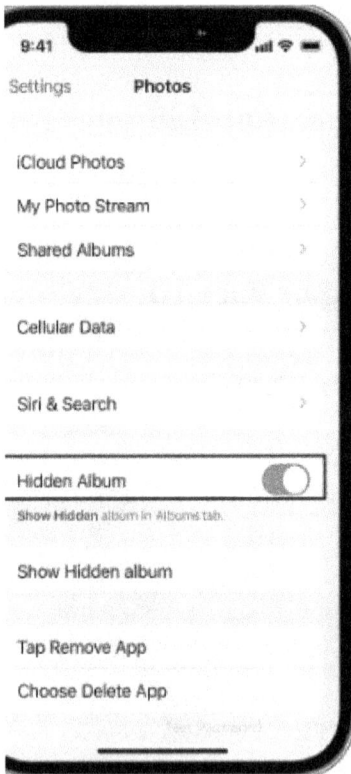

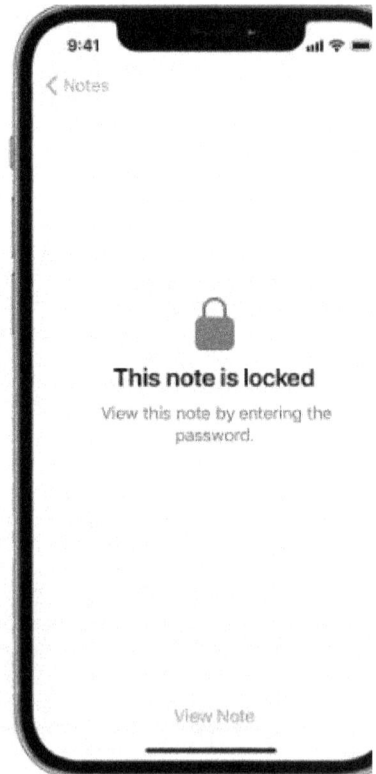

For Professionals: Built-in VPN, Hide My Email & Mail Privacy Protection

If you deal with sensitive work, Apple has **enterprise-grade privacy tools** you should know about.

Private Relay (VPN-like feature):

- Found in **Settings** → **Apple ID** → **iCloud** → **Private Relay**
- Masks your IP address and browsing activity from ISPs and advertisers
- Works only with Safari

Ideal for remote workers, journalists, and travelers

Hide My Email:

- Creates a **randomized Apple-generated email address** you can use to sign up for websites

- Emails forward to your real inbox without revealing your real address
- Enable via:
  - **Settings → Apple ID → iCloud → Hide My Email**

Mail Privacy Protection:

- Found in **Settings → Mail → Privacy Protection**
- Blocks senders from knowing when/where you open emails

Helps reduce spam and email tracking

Use all three for a total digital privacy shield—without any third-party tools.

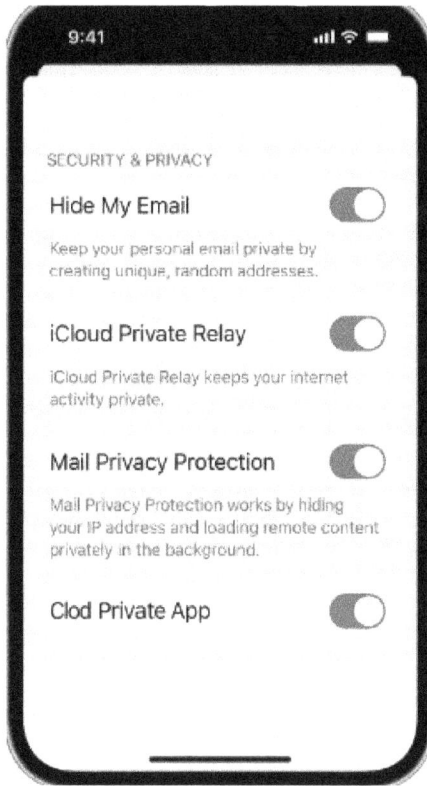

SECURITY & PRIVACY

Hide My Email

Keep your personal email private by creating unique, random addresses.

iCloud Private Relay

iCloud Private Relay keeps your internet activity private.

Mail Privacy Protection

Mail Privacy Protection works by hiding your IP address and loading remote content privately in the background.

Clod Private App

Chapter Summary

Now your iPhone doesn't just perform—it protects.

- ✓ Use Face ID, Passcode, and Lockdown Mode to secure your device

- ✓ Review and manage camera, mic, and location permissions

- ✓ Set up parental controls or simplified modes for loved ones

✓ Hide apps, notes, photos, and content with built-in tools

✓ Use Apple's native VPN, Hide My Email, and privacy protection for pro-level anonymity

Next in Chapter 10, we'll learn how to back up your data, sync your content, and use iCloud safely and efficiently.

CHAPTER 10

iCloud, Backup & Data Transfer

In a digital world, your data is everything—photos, contacts, notes, memories, files. If your iPhone is lost, stolen, or upgraded, iCloud and backup tools make sure you don't lose a thing.

In this chapter, you'll learn how to:

- Understand iCloud and what it syncs

- Back up and restore your iPhone

- Move data from your old device

- Use AirDrop, iCloud Drive, and shared photo albums

- Create a personal **Data Control Dashboard** to simplify your digital life

iCloud Basics: Storage Plans &

What's Synced

iCloud is Apple's secure cloud service that keeps your data backed up, synced, and accessible across all your devices.

What iCloud Can Sync:

- Photos & Videos
- Contacts
- Calendars
- Messages
- Notes
- Reminders
- Safari bookmarks
- Health & Wallet data
- App data and settings

Storage Plans:

Every Apple ID includes **5GB free**—but it fills up fast. To upgrade:

1. Go to Settings → **[Your Name]** → iCloud → **Manage Storage** → **Change Plan**

2. Choose a plan:

   - 50GB (great for single users)

   - 200GB (ideal for families)

   - 2TB (for heavy users or businesses)

You can share storage with family via Family Sharing.

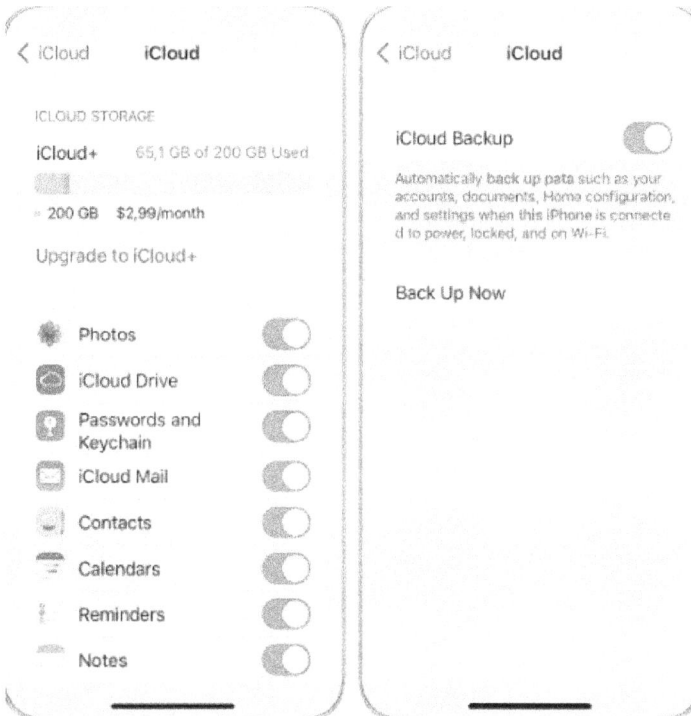

Backing Up and Restoring Your iPhone

A backup ensures that **no matter what happens**, you can restore your entire phone including messages, settings, photos, and apps.

To Turn On iCloud Backup:

1. Go to Settings → **[Your Name]** → **iCloud** → **iCloud Backup**
2. Toggle **iCloud Backup ON**
3. Tap **Back Up Now** to start instantly

iCloud will auto-backup **daily** when:

- iPhone is plugged in
- Connected to Wi-Fi
- Screen is locked

To Restore from a Backup (on a new or reset phone):

1. Follow the setup screen prompts

2. When asked, choose **Restore from iCloud Backup**

3. Sign in to your Apple ID

4. Select your most recent backup

Make sure to stay connected to Wi-Fi during restoration— it may take a while depending on size.

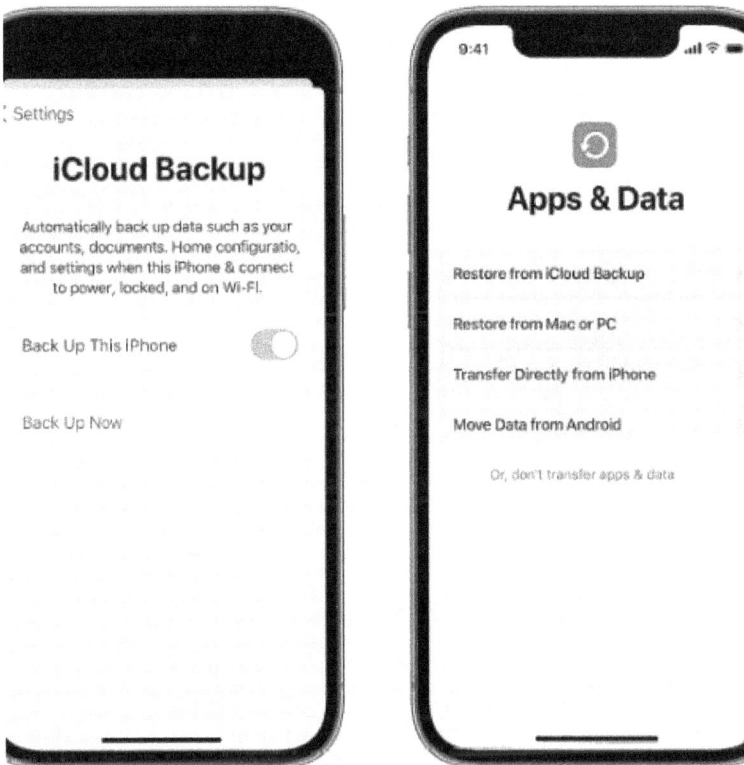

Transferring Data from Old Phone or Android

Whether you're upgrading from an old iPhone or switching from Android, Apple makes moving your data surprisingly simple.

iPhone to iPhone Transfer:

1. Turn on both phones
2. Place them side by side
3. On your new iPhone, follow the on-screen **Quick Start** steps
4. Use your old iPhone to scan the new one's animation
5. Data will transfer wirelessly or via cable

Move from Android:

1. Download the **Move to iOS app** on your Android device

2. During iPhone setup, choose **Move Data from Android**

3. Enter the code displayed on your iPhone

4. Select what to transfer (contacts, photos, messages, etc.)

Your Android data will move securely over Wi-Fi without needing a computer.

Using AirDrop, iCloud Drive & Shared Albums

AirDrop:

Quickly share photos, files, or links with nearby Apple devices.

To Use AirDrop:

- Open a photo or file
- Tap the **Share icon**
- Tap **AirDrop → Choose device**
- Accept the transfer on the other device

Great for sending vacation photos to family or sharing docs in meetings—no Wi-Fi needed.

iCloud Drive:

Apple's built-in cloud file system—think of it like Dropbox or Google Drive.

To Use:

- Open the **Files app**

- Tap Browse → **iCloud Drive**

- Save, upload, or move documents, PDFs, and app files

- Share links to files with collaborators

Shared Albums:

To privately share photo albums with friends/family:

1. Open **Photos app**

2. Tap **Albums** → + → **New Shared Album**

3. Name the album and invite people

4. Add photos → They'll get a notification to view, like, or comment

Great for family trips, school events, or collaborative photo journals.

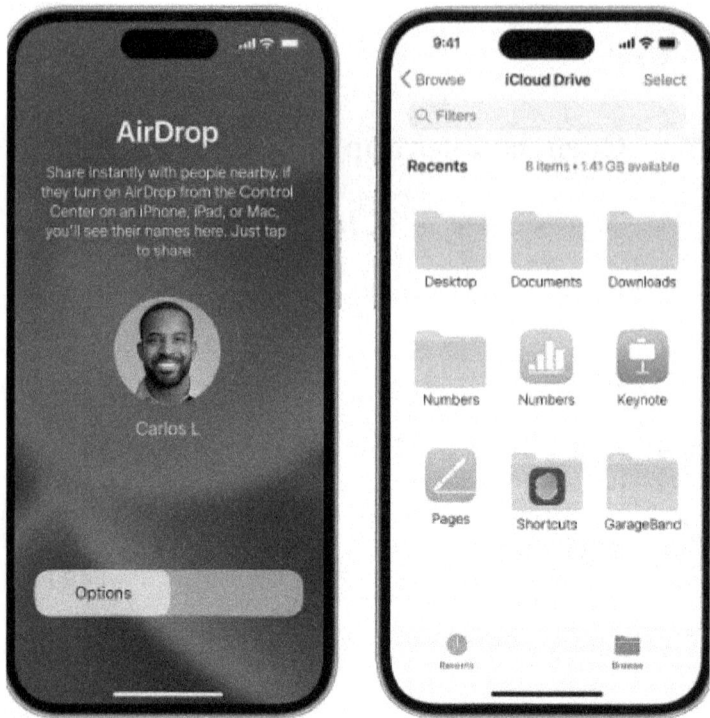

Bonus: Create a Custom "Data Control Dashboard"

Want a simpler way to manage your digital life?

Here's how to build a **Data Dashboard** using just native tools:

Use These Apps Together:

- **Reminders** → List what you back up and when

- **Notes** → Record storage plans, login details, sharing links

- **Calendar** → Monthly alert to review backups or storage

- **Shortcuts app** → One-tap action to:
  - Open Files
  - Check iCloud Storage
  - Run a manual backup
  - Open shared albums

Example Shortcut: "Data Snapshot" opens Notes, Files, and iCloud Backup screen with one tap.

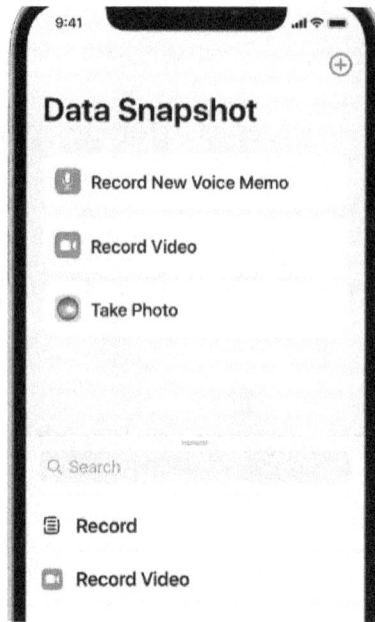

Chapter Summary

- ✓ Understand iCloud's syncing and storage plans

- ✓ Back up your iPhone and restore it safely

- ✓ Transfer data from an old iPhone or Android device

- ✓ Use AirDrop, iCloud Drive, and Shared Albums effectively

- ✓ Build a personal Data Control Dashboard for peace of mind

Coming up in Chapter 11, we'll cover Troubleshooting &

Hidden Features—from fixing common bugs to discovering Apple's secret tricks and settings.

CHAPTER 11

Troubleshooting & Hidden Features

Even the smartest iPhone isn't perfect every day. Apps may freeze. Screens may lag. Battery may drain faster than expected.

This chapter equips you with **practical solutions** to everyday issues and reveals Apple's **best-kept secrets** that most users don't even know exist.

Let's make sure your iPhone 16 runs smoother, faster, and smarter.

Common Issues and Quick Fixes

Below are some of the most frequent problems and how to solve them in seconds.

Frozen or Unresponsive Screen

Fix: Perform a *soft reset*

1. Press and release **Volume Up**

2. Press and release **Volume Down**

3. Press and hold **Side Button** until the Apple logo appears

Phone Running Slowly

- Close unused apps (swipe up and pause → swipe away)

- Restart your iPhone

- Delete unused apps or offload app data in **Settings → iPhone Storage**

App Not Opening or Crashing

- Force quit the app → Reopen

- Update the app via App Store

- Delete and reinstall the app

Wi-Fi or Bluetooth Not Connecting

137

- Toggle **Airplane Mode ON** → **then OFF**

- Go to **Settings** → **Wi-Fi** → Forget network → Reconnect

- Restart your iPhone

Battery Draining Fast

- See "Battery Optimization" at end of this chapter

COMMON ISSUES AND QUICK FIXES

	FROZEN/ UNRESPONSIVE SCREEN	Perform a soft reset
	PHONE RUNNING SLOWLY	Close unused apps or restart phone
	APP NOT OPENING OR CRASHING	Force quit or update the app
	WI-FI OR BLUETOOTH NOT CONNECTING	Toggle airplane mode or restart phone
ON	airplane mode or restart phone	Delete App

Reset Options (Soft Reset vs. Factory Reset)

Resetting can fix deeper problems but there's a right way and a risky way to do it.

Soft Reset (Safe, No Data Loss)

Just like restarting a computer:

- Press Volume Up
- Press Volume Down
- Hold Side Button until Apple logo appears

Factory Reset (Erases Everything!)

Use only when you're:

- Selling or giving away your phone
- Starting fresh due to serious errors
- Restoring from backup

To Do a Factory Reset:

1. Go to **Settings → General → Transfer or Reset iPhone**

2. Tap **Erase All Content and Settings**

Make sure to back up first. This will delete all your data.

Hidden Tips & Tricks

Let's uncover the **secret moves and features** that even long-time users often miss.

Back Tap Actions

Your iPhone's back panel can become a custom shortcut button.

To Set Up:

1. Go to **Settings → Accessibility → Touch → Back Tap**

2. Choose:

   • Double **Tap or Triple Tap**

- Assign actions like:

    o Take a screenshot

    o Open Control Center

    o Launch a Shortcut

    o Mute volume

    o Open flashlight

Try "Double Tap = Camera" for instant access.

Sound Recognition

iPhone can **listen for important sounds** and notify you—great for hearing impaired users or those in loud environments.

To Enable:

1. Go to Settings → **Accessibility** → **Sound Recognition**

2. Turn it ON → Choose sounds like:

    - Smoke Alarm

    - Doorbell

- Baby Crying

- Shouting

- Water Running

You'll get a **vibration + notification** when the sound is detected.

Live Voicemail (iOS 18 Feature)

See real-time **transcripts of voicemails** as they're being left—like call screening.

To Enable:

1. Go to **Settings → Phone → Live Voicemail**
2. Toggle ON

When someone leaves a voicemail, you'll see their words appear on-screen **as they speak,** and you can **choose to pick up mid-message**.

Think of it like voicemail meets live captions—super helpful for screening unknown calls.

Battery Optimization Tips

Don't let your battery drain faster than your energy. Try these tips:

✅ **Check What's Draining Battery:**

- Go to **Settings → Battery**
- Scroll down to see apps and their battery usage

✅ **Enable Low Power Mode:**

- Settings → Battery → Turn ON
- Or add to Control Center for one-tap access

✅ **Turn Off Background App Refresh:**

- Settings → General → Background App Refresh → Off or Wi-Fi Only

✅ **Auto-Lock Your Screen:**

- Settings → Display & Brightness → Auto-Lock → 30 seconds or 1 min

✅ **Disable Location Services for Unused Apps:**

- Settings → Privacy → Location Services

☑ Use Dark Mode:

- Saves battery on OLED screens (like iPhone 16)

Battery health is best maintained by keeping your charge between 30–80% long term.

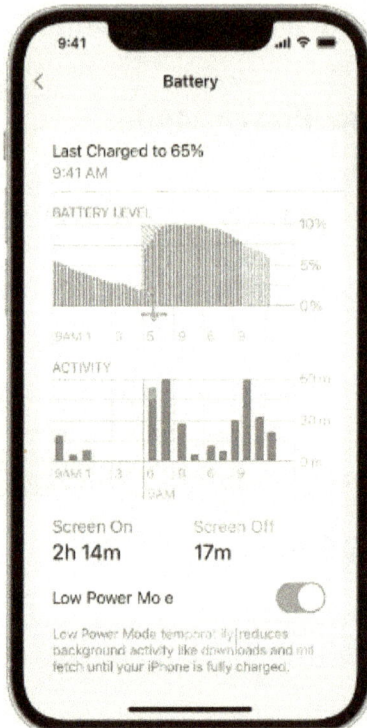

Chapter Summary

Now you're armed with solutions and secrets most users

never discover:

- ✓ Fix common glitches and frozen screens
- ✓ Understand soft vs factory reset—when and how to use
- ✓ Use Back Tap, Sound Recognition, and Live Voicemail
- ✓ Optimize your battery for longer life
- ✓ Add shortcuts, alerts, and real-time visibility to your device

In Chapter 12, we'll look at Expert Tips, iOS Hacks, and Recommended Apps—so you can go from confident to advanced in just a few swipes.

CHAPTER 12

Expert Tips, iOS Hacks & Recommended Apps

You've mastered the basics. Now it's time to unlock the *next level*.

In this final chapter, we'll dive into:

- Must-have utilities that supercharge daily use

- Pro-grade photography apps for stunning photos

- Hidden iOS settings that quietly boost performance

- Apple's top accessibility tools—designed for *everyone*

Best iOS 18 Utilities & Apps for Daily Use

Here's a hand-picked list of apps and iOS tools that can make your daily life smoother, more productive, and surprisingly fun.

1. Files App (Built-in)

- Use it as a mini hard drive
- Browse iCloud Drive, On My iPhone, or USB drives
- Scan documents directly into folders

2. Notes App + Scanner

- Scan paper docs into searchable text
- Lock private notes with Face ID
- Sync across iPad, Mac, and iCloud

Tip: Use folders and tags in Notes to organize projects like a digital filing cabinet.

3. Shortcuts App

- Automate routines like:
  - "Morning Prep" → Weather + Calendar + Music

- "Drive Mode" → Open Maps + DND + Spotify

- Run via widgets or Siri

4. Calculator with Scientific Mode

- Rotate your iPhone sideways in Calculator for advanced functions

- Great for students, engineers, or budgeting

5. Cleanfox / Email Cleanup Apps

- Unsubscribe from junk mail in seconds

- Keep inbox zero with one-tap declutter

6. Widgetsmith

- Design custom widgets for time, calendar, quotes, battery, reminders

- Personalize your Home Screen aesthetic

Combine with Focus modes for a themed digital lifestyle.

Pro iPhone Photography Apps

The iPhone 16 already has an amazing camera. But if you want **more control, creative effects, or pro-level edits**, these apps take it further:

1. Halide Mark II

- Full manual camera control (ISO, shutter speed, RAW photos)
- Ideal for pros or photography students

2. Lightroom Mobile

- Advanced photo editing with presets, exposure control, color grading
- Syncs with Adobe Cloud

3. Snapseed

- Free, intuitive, powerful edits
- Selective editing, double exposure, healing tool

4. Focus

- Create DSLR-like bokeh even after taking the shot

- Adjust focus depth with AI

5. Canva

- Great for social media content, thumbnails, or event invites

- Drag-and-drop design with templates

Want better Instagram? Combine Halide + Lightroom + Canva.

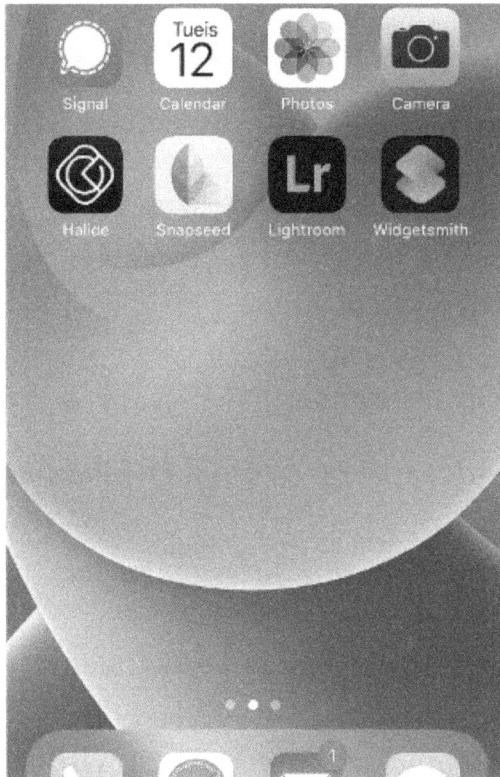

Hidden iOS Settings That Boost Performance

These secret gems are often overlooked but they can **speed up your phone, boost battery, and tighten privacy**.

Limit Background Activity

- Go to Settings → **General** → **Background App Refresh** → **Off** or Wi-Fi only
- Reduces battery and data usage

Turn Off System Ads

- Go to **Settings** → **Privacy & Security** → **Apple Advertising** → **Turn Off Personalized Ads**

Stop Apple from profiling you for ads

Offload Unused Apps (Auto-Clean)

- Settings → App Store → **Offload Unused Apps: ON**

Saves space while keeping your data intact

Limit Motion & Visual Effects

- Accessibility → Motion → **Reduce Motion: ON**

Speeds up animation + reduces strain on the eyes

Analytics & Diagnostics

- Settings → Privacy & Security → Analytics & Improvements
- Turn off "Share iPhone Analytics" and "Improve Siri & Dictation"

These tweaks save battery, reduce lag, and improve privacy in the background.

Accessibility Tools for Everyone

Apple's accessibility features aren't just for users with disabilities, they make the iPhone easier and more personalized for everyone.

Voice Control

- Control the iPhone entirely with your voice
- Navigate, open apps, tap buttons, and more

Magnifier

- Turns your camera into a digital magnifying glass

- Great for reading menus, product labels, or fine print

Sound Recognition

- Get alerts when doorbells, alarms, or babies cry
- See Chapter 11 for full breakdown

Guided Access

- Lock the iPhone to one **app only**—great for:
  - Kids
  - Seniors
  - Public demo use

To Enable:

1. Settings → Accessibility → Guided Access
2. Triple-click Side Button in any app to lock it
3. Set a passcode to exit

Display Enhancements

- Bold Text

- Larger Text

- Color Filters

- Invert Colors

- Reduce Transparency

Senior Tip: Combine Magnifier, Bold Text, and Voice Feedback for the ultimate clarity experience.

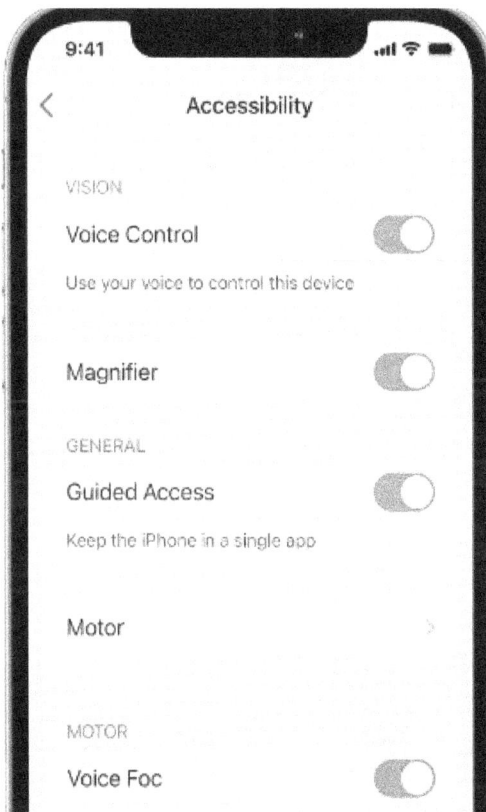

Chapter Summary

Let's wrap up with your final level-up checklist:

- ✓ Use built-in iOS apps and widgets to simplify daily life
- ✓ Install pro photography apps for better control and creativity
- ✓ Enable hidden iOS settings that enhance speed, privacy, and battery
- ✓ Customize your experience with powerful Accessibility tools

You're now more than just an iPhone user—you're in control, informed, and fully equipped to use your iPhone 16 like a pro.

FINAL THOUGHT

From unboxing to automation, from FaceTime to security, this guide was built to empower you with confidence. Keep exploring. Keep learning. Your iPhone 16 is more powerful than ever—and now, so are you.

Glossary of iPhone Terms

Term	Definition
AirDrop	A wireless way to share files, photos, or links between Apple devices
App Library	A categorized app organizer that appears after your last Home Screen page
Control Center	A pull-down panel to quickly access brightness, Wi-Fi, volume, etc.
Face ID	Face recognition system to unlock iPhone or approve purchases

Haptic Feedback	A small vibration you feel when tapping the screen
iCloud	Apple's cloud storage system for backing up data and syncing devices
Live Voicemail	Shows a live transcript of voicemails as they're being recorded
Lock Screen	The screen shown before unlocking your iPhone
Notification Center	The panel showing messages, alerts, and updates
Shortcuts	A built-in app that lets you automate sequences of

tasks with a single tap

Siri	Apple's voice assistant for hands-free control of your iPhone
StandBy	Turns your iPhone into a smart display when charging and idle
Widgets	At-a-glance tools on the Home or Lock Screen that show app info

Icon Reference Guide

Icon	**Meaning**

▭ Battery	Indicates remaining battery level
▾ıl Signal Bars	Shows mobile network signal strength
▯ Vibrate Mode	Your phone is in silent/vibration mode
🔒 Lock	iPhone is locked
⊕ Globe	Change keyboard language
🕓 Clock	Indicates scheduled alarm or timer is active
🔔 Bell w/ Slash	Do Not Disturb or Focus Mode is ON
↻ Circle Arrows	Syncing or updating content

☁☐ Cloud w/ Arrow	iCloud download pending or in progress
🔲 Rectangle w/ Arrow	Screen mirroring active

If you ever see an unfamiliar icon, swipe down for Control Center or tap and hold it for a label.

Keyboard Shortcuts (External Bluetooth Keyboard)

If you connect a wireless keyboard to your iPhone, you can use these handy shortcuts:

Shortcut	Action
⌘ + Spacebar	Open Spotlight Search
⌘ + H	Go to Home Screen
⌘ + Tab	Switch between recent

	apps
⌘ + Shift + 3	Take a screenshot
⌘ + R	Refresh in Safari
⌘ + W	Close tab in Safari
⌘ + T	New tab in Safari
⌘ + L	Focus address bar in Safari
⌘ + A / C / V / X / Z	Select All / Copy / Paste / Cut / Undo (universal)

Holding down the Command (⌘) key shows a list of available shortcuts in any app.

iOS 18 Updates Tracker

As iOS evolves, features improve. Keep an eye on version updates.

How to Check for Updates:

- Go to Settings → General → Software Update

- Tap Download and Install if available

What iOS 18 Updates Usually Include:

- Bug fixes and battery improvements

- New features (e.g., added widgets, enhanced privacy tools)

- Security patches

- UI enhancements or added Siri abilities

Always update when connected to Wi-Fi with enough battery.

Acknowledgments

To every beginner, senior, and curious learner who picked up this book—thank you. Your desire to grow, adapt, and explore the digital world inspires everything I write.

Special thanks to my family and friends for their encouragement, to my team for their unwavering support, and to every readers whose feedback helped guide this book into something truly useful.